On The Occurrence Of Mammoth And Mastodon Remains Around Hudson Bay

Robert Bell

BULLETIN OF THE GEOLOGICAL SOCIETY OF AMERICA

VOL. 9, PP. 369-390

ON THE OCCURRENCE OF MAMMOTH AND MASTODON REMAINS AROUND HUDSON BAY

BY

ROBERT BELL

ROCHESTER

PUBLISHED BY THE SOCIETY

JUNE, 1898

BULLETIN OF THE GEOLOGICAL SOCIETY OF AMERICA

VOL. 9, PP. 369-390   JUNE 22, 1898

ON THE OCCURRENCE OF MAMMOTH AND MASTODON REMAINS AROUND HUDSON BAY

BY ROBERT BELL

(Read before the Society December 29, 1897)

CONTENTS

INTRODUCTION

In the following paper on the discovery of the remains of both the mammoth and the mastodon around Hudson bay it is proposed to give a short account of the geological surroundings of each of these discoveries, and to refer very briefly to the consequent extension of our knowledge of the geographical range of both species over a vast area, and also to notice some of the questions which they suggest, such as those relating to the climates of the regions in which these animals flourished, the geological dates when they lived in the old and the new worlds, their general geographical distribution, migrations, causes of their extinction, and other matters of geological importance concerning them.

THE OCCURRENCE OF A MAMMOTH TOOTH ON THE EAST COAST OF HUDSON BAY

In 1877 I made a geological exploration of the east side of Hudson bay. In the following year a remarkable molar tooth of a mammoth was picked up on the rocky surface of Long island and sent to me. This island is narrow, thirty miles in length, and lies near the Eastmain coast, its southwestern extremity being just north of cape Jones, which is the point where James bay opens out into Hudson bay proper. It is composed of almost bare rock and has a ridge of basalt running down its center. Although it is south of the limit of timber on the mainland, no trees grow upon the island itself. At the time when mammoths lived upon the island it must have been more or less wooded in order to furnish food for them, and, from our knowledge of the uprising of the land in this region in Pleistocene times, it probably stood at a lower level, and would therefore be of smaller size.

Photographs showing this tooth in different positions were sent to Professor W. Boyd Dawkins, who had made a special study of fossil elephants, and to Professor E. D. Cope, with requests for their decisions as to the species to which it belonged. The following are their replies:

" WOODHURST, FALLOWFIELD, MANCHESTER, *10th July, 1882.*

"The very interesting specimen is a worn stump of an upper molar belonging to *Elephas columbi*, an animal which, in my belief, is one of the varieties of the stock from which the mammoth, Indian elephant, and *E. armeniacus* have been derived. The stump is wonderfully like some of the mammoth's in my collection, but it is narrower.

"(Signed) W. BOYD DAWKINS."

"PHILADELPHIA, *Dec. 19th, 1882.*

"I have never seen a tooth which presents all the peculiarities of this one, but each of its characters can be found separate in different teeth of the mammoth. It is probably a last deciduous molar of a variety between the typical *E. primigenius* and the smooth-plated *E. columbi.*

"(Signed) E. D. COPE."

Elephas columbi of Dr Hugh Falconer, to which this molar belongs, according to Professors W. Boyd Dawkins and E. D. Cope, has been found on the Pacific coast of Alaska.* Falconer only knew of its remains in

FIGURE 1.—*Molar of Elephas columbi from Long Island, Hudson Bay.* (⅓ natural size.)

the more southern of the United States and Mexico; but the present discovery, and that of a similar molar, near Edmonton, N. W. T., taken in connection with its occurrence in Alaska, shows that its range in North America was even more extensive than that of *E. primigenius.* Considering how very rare the discovery of elephantine remains of any kind has hitherto been over all that great portion of the continent between Bering strait and the vicinity of lake Erie, we may reasonably expect that among

---

* Bulletin of the U. S. Geol. Survey, no. 84, 1892.

future discoveries of such remains in this vast interval some at least will prove to belong to *E. columbi.*

Both *E. columbi* and *E. primigenius* may have retreated southward in America on account of the increasing severity of the climate, as did the moose and the bison, or without any sufficient change in this respect, having become accustomed to the better climate which they found to the south, they abandoned the northern regions altogether.   It is to be noted that the invasion of North America by mammoths was from north to south, or in the opposite direction from the invasion of Siberia, and this circumstance may have made an important difference as to the character of their subsequent movements on the two continents and as to the latitudes in which they survived the longest in the respective regions.

Dr Hugh Falconer, who, with ample materials at his command, gave the subject very careful study, regards this elephant not as a variety of *E. primigenius,* but as belonging to a decidedly different species.   Its occurrence on Hudson bay is of much interest and opens up various questions, not only as to the geographical range of the species, but also in regard to former climatic conditions of that region and the distribution of land and water at a comparatively recent period which would result from the former levels of different parts of the continent.

The accompanying figure (1) will enable those who are critically versed in such matters to judge for themselves as to the species of mammoth to which this molar belonged.

MAMMOTH TOOTH FOUND NEAR EDMONTON

Two or three years ago an incomplete molar, which appears to belong to *Elephas columbi,* was found in the superficial deposits in one of the banks of the North Saskatchewan river, about six miles above Edmonton, but no particulars in reference to the discovery are available.   The specimen is in the museum of the Geological Survey at Ottawa.   The Saskatchewan at Edmonton has an altitude of about 2,200 feet above the sea.   The surrounding district is a fine agricultural country, with a deep covering of till, overlaid in parts with stratified sands, gravels, and clays, all having a rolling surface.

ELEPHANT REMAINS IN THE FAR NORTHWEST

It has long been known that mammoth remains exist in several places in the far northwestern regions of North America.   Dall mentions the occurrence of bones and tusks of these animals on the lower Yukon river, and a few years ago some of their remains were purchased for the museum of the Geological Survey at Ottawa from Mr F. Mercier, who had brought them from some place on this river within the Alaskan boundary. The late Mr Robert Campbell, of the Hudson's Bay Company's service, with whom I have conversed on the subject, told me that he had found elephant bones in a river bank near the junction of the Lewis and Pelly to form the Yukon. One of these bones was sent to England and identified by Sir John Richardson as belonging to *Elephas primigenius* (Blumenbach). The late Mr Richard Hardisty and the officers of the Hudson's Bay Company have informed me that in passing along Rat river, a small tributary of the McKenzie on the west side of its delta, the Bell river, a branch of the Porcupine, and along this stream itself they have seen in various places bones of elephants projecting from the clay or other superficial deposits forming the banks. Mr Hardisty said that in the vicinity of these rivers he had frequently seen skulls of the musk-ox (often called " buffalo " in the far north) lying on the surface of the ground, mostly in swamps and partly covered with moss. This animal is not now found living west of the McKenzie river, although it is quite common over the great region to the eastward of it as far as Hudson bay and thence across the large islands lying to the northeastward, which carry its range to northern Greenland. Its desertion of the country west of the McKenzie river is one of those instances of the long-period or the final regional migrations of the larger mammals which have not yet been satisfactorily accounted for.

The late venerable Archdeacon R. McDonald, whom the writer has had the pleasure of meeting in the McKenzie valley, some years ago presented to the British Museum bones of the mammoth, the horse, and the musk-ox. I have obtained from York factory, on the west side of Hudson bay, through the kindness of Dr Percy Mathews, part of the skull of a horse which was found there half embedded in the soil ; but this discovery may have no geological significance, as it possibly belonged to a domestic horse, although I could not hear of any of these animals having ever been landed at this place, although cattle for the Red River settlement and for local use have been imported to this establishment from England.

Mr William Ogilvie obtained during the present year two horns of the existing bison (which I have seen) from the auriferous gravels of Bonanza and Eldorado creeks, in the Klondike district of the Yukon near the intersection of the one hundred and forty-first meridian. From the same gravels he also obtained some coniferous wood and part of a skull and part of an antler of the reindeer. The range of the bison in modern times has not come within several hundred miles of this district.

GEOLOGICAL HISTORY OF THE MASTODON AND MAMMOTH

The elephant family made its first appearance in the Miocene period in southeastern Asia. The earliest of the true Proboscidea were (1) the Stegodons, which were the ancestors of the mastodons, the mammoths, and the Indian elephant, and (2) the Loxodons, the ancestors of certain fossil elephants of Europe and also of the African elephant. As time went on and new species appeared, the elephants spread from their original birthplace into Europe, Africa, northeastern Asia, and thence into America over a neck of land which at a comparatively recent geological period closed up Bering strait. That such a land connection existed and that mammoths passed over it appears to be proven from the fact that remains of these animals have been found on Saint George and Saint Paul islands, of the Pribilof group, and on Unalaska, one of the Aleutian islands. The deepest part of Bering strait is covered by only 300 feet of water, and, since a subsidence of more than this amount has taken place in the Pleistocene period, there is no doubt that an isthmus connected the two continents at no distant date, and that men as well as mammoths and other animals may have walked over it. In the old world, mastodons died about the close of the Pliocene, but the American species (*Mastodon americanus*) lived on, along with the mammoth, into the human period. Falconer says : *

"Commencing with the older strata of the sub-Apennines and of the Val d'Arno and ascending to the superficial gravels or quaternary deposits of comparatively modern origin, at least four well defined species of fossil elephants have been ascertained to have existed in Europe, namely, *Elephas* (*Loxodon*) *meridionalis*, *E. antiquus*, *E. primigenius*, and *E.* (*Loxodon*) *africanus fossilis*."

A little further on the same writer says : †

"If the asserted facts be correct, they seem clearly to indicate that the older elephants of Europe, such as *E. meridionalis* and *E. antiquus*, were not the stocks from which the later species, *E. primigenius* and *E. africanus*, sprung, and that we must look elsewhere for their origin. The nearest affinity, and that a very close one, of the European *E. meridionalis* is with the Miocene *E.* (*Loxodon*) *planifrons* of India, and of *E. primigenius* with the existing Indian species."

Again Falconer writes : ‡

"The result of any observation is that the ancient mammoth of the preglacial

---

* Palæontological memoirs and notes of the late Hugh Falconer, A. M., M. D., vol. ii, p. 251.
† Op. cit., p. 254.
‡ Op. cit., p. 252.

'forest bed' of the Norfolk coast differs less from the later form, occurring on the banks of the Lena, than does the latter from the comparatively modern mammoth of the superficial bogs of North America, which I regard as being only a slight geographical variety of the same species."

THE EXTINCTION OF SPECIFIC FORMS OF MAMMALS

From the first appearance of Proboscideans to the Recent period, one form after another has passed away, to be succeeded by another, until we have arrived at the immediate precursor of the existing Indian elephant, which appears to be specifically identical with the mammoth. Falconer insists on the importance of the fact that throughout the whole geological history of each species of elephant there is great persistence in the structure and mode of growth of each of the teeth, and that this is the best single character by which to distinguish one species from another. He finds, after a critical examination of a great number of specimens, that in the mammoth each of the molars is subject to the same history and the same variation as the corresponding molar in the living Indian elephant.* Even if zoologists agree that these two elephants belong to the same species, *E. primigenius* is sufficiently distinguished as a well marked variety to deserve recognition for all the purposes of geological description. No single cause may account for the extinction of the mammoth all over the world. As will be pointed out further on, it may have been due to the climatic changes in Siberia, while human agency may have been the final cause in Europe and North America; but whether the Indian elephant is specifically identical with *E. primigenius* or not, there appears to be at the present time a general tendency to extinction in the existing form, as one which has run its course. The cause of this is not apparent, unless it be owing to the well known general law that the higher species of animals have a shorter term of existence than the lower ones, and that the period of their survival is somewhat proportionate to their rank in the scale of being.

The history of the larger mammals shows that when the geographical range of a species has become greatly diminished, with a corresponding reduction in its numbers, it does not recover lost ground, but hastens to its end. These conditions now apply to the Indian elephant, whether he represents the very circumscribed remnant of the once almost cosmopolitan mammoth or not. Another sign of the approaching extinction of this species is its loss of reproductive vigor, as evidenced by the fact that it will scarcely breed at all in the state of domestication.

As the mammoth lived contemporaneously in the old and new worlds

---

* Op. cit., p. 168.

after the Glacial period, there could scarcely have been a sudden change
in climate or conditions which would account for its disappearance in
both hemispheres about the same time.   The Indian species maintains
its existence in the original home of the whole race because the conditions favorable to Stegodont elephant life probably continue to be better
there than anywhere else.

MIGRATIONS OF NORTHERN MAMMALS

Popular writers on this subject appear to associate the existence of
entire carcasses of mammoths about the mouth of the Lena river with
the extinction of the species all over Europe, Asia, and North America,
whereas this fact is only a local circumstance in the long history of the
animal.

The migration of birds and mammals, which is so characteristic of
many species at the present day, has been going on for ages.   The alter-
nation of the seasons in the northern hemisphere would naturally stimu-
late and develop a tendency among such creatures to move northward
and southward with the changing temperature and food supply, and the
elephants would be no exception.   The reindeer, with whose bones those
of the mammoth are associated in Europe and Asia, retains its migratory
instincts in both the old world and the new.

But the woodland variety of this species (called the caribou in Canada)
is not migratory, and it is possible and even probable that there were
also migratory and non-migratory mammoths, according as they in
habited (like the reindeer) the open northern barren lands or the more
southern wooded country in either the old or the new world.   The
musk-ox and the American bison made extensive annual migrations
The Arctic fox travels hundreds of miles north and south every year with
the change of the seasons.   The Canada lynx is one of the most migratory
of North American mammals, but its movements are governed by food
supply alone, and depend upon the varying abundance or scarcity of its
principal prey, the common American hare.   Even the little lemming
perform wonderful migrations, impelled, as it were, by an irresistible
impulse.

The moose or American elk (*Alces americanus*) migrates slowly from
one large area to another through periods extending over many years
For example, in the Gaspé peninsula the last interval between its leaving
and again returning to the same district was upward of half a century
and in the region between the upper Great Lakes and James bay the period
between his last withdrawal and reappearance has been still longer
Within the historic period the bison roamed as far east as lake Superior

and lake Erie, and in modern geological times it ranged into the distant
northwest as far as the Yukon river, where, last year, Mr William Ogilvie
obtained two of its horns (which I have seen) in the gold-bearing gravel
of Bonanza creek, in latitude 64°. The biche or red deer (*Cervus cana-
densis* Erxleben) inhabited the St. Lawrence valley eastward to the outlet
of lake Ontario in comparatively recent times, its remains in a good state
of preservation having been found embedded in shell marl in at least
two localities near Kingston, and also in the sand and gravel of Burlington Heights at a depth of 30 feet from the surface and at a height of
77 feet above lake Ontario. Thirty years ago it was common in eastern
Manitoba, but now it has retreated still further west. Such examples as
the foregoing of annual and long-period migrations favor the supposition
that the mammoth, in addition to its slower general movement of dis-
persion to remote parts of the earth, made annual migrations in regions
where such movements would be beneficial to him and might naturally
be expected, as in northern Siberia. The average distance at the present
day between the verge of the forest in northern Europe and Siberia and
the coast of the Arctic sea is from 100 to 250 miles.

Even if the distance had been as great as this at the time when the
mammoth inhabited these regions, which is not probable, this animal
could easily move from the forest to the sea coast and back again between
spring and autumn. But there is evidence that the modern forest-line
has been retreating southward in both the old and new worlds. This
tendency has prevailed for a great length of time, as is shown by the
remains of trees of existing species on the coasts of Bering strait beyond
the present limits of timber, and in Melville island off the northern coast
of this continent, the latter occurring between 500 and 600 miles directly
north of the nearest trees now growing on the Coppermine river and near
Great Bear lake. The increased severity of the seasons in Greenland in
historic times is another evidence of the deterioration of the subarctic
climate, which appears to have this tendency all round the world, with
perhaps a few local exceptions, as in one part of northwestern Alaska,
due probably to a favorable change in the ocean current there.

Extinction of the Mammoth in Siberia

The mammoth in northern Siberia probably passed the winters within
the forest-line, where he would find shelter from the chilling winds and
where he might live well, browsing on the small branchy spruce, larch,
birch, etcetera. With the advent of spring he would begin his northward march, taking advantage of the long daylight, and he would spend

part of the summer and the autumn roving about the shore of the Arctic sea, enjoying the cool weather and finding abundant sustenance on the small trees and the alder, willow, and birch brushwood. Then, with the beginning of the severe weather, he would turn his footsteps toward his winter quarters and move south as the season advanced. The periods of their annual migrations having become settled, it would be difficult or impossible to overcome the inertia of long-fixed habit, and they would be obliged to endure the increasing severity of the climate on the borders of the Arctic sea. In the meantime their numbers would be greatly diminished from causes to be mentioned further on. At length, those which journeyed as far as the sea coast might be reduced to the single herd which migrated to the mouth of the Lena, where the climate of autumn would be the best on the coast, owing to the large quantity of warm water from the south which accumulates off the mouth of this great river.

At this stage, if an unusually early and severe season were to set in, accompanied by great snow-storms, before the herd had started for the south, the result might be disastrous to the remaining mammoths. The now stunted brush would be covered by the deep snow, on which perhaps a strong crust had formed, thus preventing the animals from obtaining any food, while the almost continuous darkness of the early winter would also operate against them. The same conditions would make it difficult or impossible for them to travel. Other individuals or herds which did not migrate so far north may have perished from a similar cause in various parts of the region. We know how completely helpless the deer of any species become in our northern woods when caught in deep snow with a crust upon it.

Under circumstances like these the last of the mammoths would soon perish, since creatures of their organization, living upon such slightly nutritious food, must have it continuously and in large quantities. That such a process of starvation is not imaginary, I may mention the fact that the reindeer sometimes perish over large areas in our northern barren lands from this cause. Their lives depend upon a continuous supply of the reindeer-lichen, which they obtain by removing the snow or by finding the plant where the ground has been left bare by drifting. A striking instance of this occurred many years ago on Akpatok island, in Ungava bay. This large island had always swarmed with reindeer, but one winter, when the snow was deeper than usual, rain fell upon it (an almost unprecedented occurrence) and formed a heavy and permanent crust over both the bare ground and the snow, thus preventing the deer from obtaining their food. The consequence was that the whole number perished, and the island has never been restocked. If this former great herd had comprised the whole species then living, the reindeer would now be extinct.

PRESERVATION OF THE FLESH OF MAMMOTHS IN SIBERIA

The preservation till the present day of the flesh of some of the mammoths which perished in the region about the mouth of the Lena river and elsewhere proves that the carcasses must have become frozen immediately after death, and this circumstance may be accounted for in the following way: If the last of these creatures succumbed in the manner supposed, there may have been at that time a series of unusually cold years, as sometimes happens in high latitudes, and this, together with the increasing severity of the climate in general ever since, would account for the preservation of some of their carcasses in the snow and ice which have persisted in that region till the present time.

The occurrence of large numbers of the remains of mammoths in the alluvial deposits about the mouth of the Lena and other rivers may be explained by the supposition that the animals had broken through the too thin ice in attempting to cross the streams upon it on their southward migration in the autumn, and that their bodies had subsequently floated down to the still water. Indeed, it is highly probable that whole herds of these animals lost their lives in this manner. While the bison was abundant in our northwest territories it was a matter of common occurrence for large numbers of them to be drowned when attempting to cross the streams in compact droves before the ice was strong enough to bear the strain. The great abundance of bison bones in some of the fluviatile deposits in this region is easily accounted for in this way.

The mammoths, owing to their great weight, would be still more liable to such an accident. Professor Richard Lydekker, in " The Royal Natural History," lately published, speaking of the trade in ivory from Siberia, says that within a recent period, covering twenty years, 20,000 mammoths must have been discovered in that region.

IMPROBABLE THEORIES

The supposition that the mammoths of northern Siberia were frozen where we find them by a sudden change from a warm to a very cold climate, and which has remained permanently so, is as untenable as the other theory, which supposes the bones and tusks found there to be those of mammoths which were drowned in great numbers and at the same time within a limited area by a sudden cataclysm. If it were possible (which it is not) that such an abrupt change of climate could happen, it would require to be general around a great part of the globe, and there is no evidence that such a thing occurred at any time in the history of the earth. Again, to invoke the agency of sudden cataclysms to account for geological phenomena is an exploded notion which does not require discussion.

FOOD AND GEOGRAPHICAL RANGE OF THE MAMMOTH

From the remains of food found with the teeth and skeletons of the mastodon and mammoth, it has been pretty satisfactorily ascertained that in North America both of these animals subsisted largely on the twigs and boughs of northern trees, such as the spruces (*Picea*) and white cedar (*Thuja occidentalis*), together, probably, with those of other northern trees and bushes, and no doubt the food of the Siberian mammoth was of the same nature. Their large grinders and powerful muscles were admirably adapted to reduce such materials to a pulp. Both the African and Indian elephants are "coarse feeders," living principally upon the branches and bark of trees and bushes, and the mammoth, wherever he wandered, would require to subsist upon such kinds of food of this description as the country he might be in produced.

"We further know that when the mammoth pastured along the margins of the great swamps of Ohio and Kentucky the vegetation then was nearly identical with what it is now, being very different from that of Siberia" (Hugh Falconer). The same writer,* referring to *Elephas primigenius*, says of it: "A scope in space and time, taken together, has been assigned without a parallel, I believe, within the whole range of the mammalia, fossil or recent. D'Archiac, in his excellent 'Histoire des Progrès,' so late as 1848, gives a brief summary of the localites in which the remains of the mammoth (*E. primigenius*) have been said to occur, namely, from the British islands across the whole of the temperate zone of Europe and Asia and along all the coasts and islands of the Icy sea as far as the frozen cliffs of the east coast of Bering strait, in Eschscholtz bay, in Russian America as high as 66° of north latitude, over most of the United States of America, in the great valley of the Mississippi, and along the coasts of the gulf of Mexico. De Blainville, going a step beyond most of the paleontologists, doubtingly referred the fossil remains of elephants found so abundantly in tropical India to the same species, thus assigning at least half of the habitable globe for the pasture ground of the mammoth."

WOOLLY COAT OF THE SIBERIAN MAMMOTH

The wool and long hair found upon the Siberian mammoths prove

---

* Palæontological memoirs and notes of the late Hugh Falconer. London, 1868, vol. ii, p. 77.

that they had been accustomed for a great length of time to a severe climate. Although the Indian elephant inhabits a warm country, it is a well known fact that he is intolerant of great heat and suffers much when exposed to the direct rays of the sun in that climate. In the wild state he seeks the cool shade and wanders about at night or in the early morning.

Notwithstanding the heat of the climate of India, it has lately been discovered that the elephant of that country retains traces of wool like that which formerly clothed the mammoth. The presence or absence of wool or of a thicker or thinner coat of hair or fur on a mammal does not often constitute an important specific character. On the highlands of Tibet, where the climate is cold in winter, the domestic goat and the mastiff dog produce fine wool under their hair. In Canada we have examples of the same kind of growth in at least two of our common mammals, the moose and the porcupine. In the country on the south side of the Saint Lawrence, below Quebec, I have seen quantities of very fine brown wool taken from beneath the hair of moose killed in the middle of winter, which the French Canadian women were manufacturing into stockings and mittens of a superior quality. The porcupine ranges far north, and in the region west of Hudson bay he is covered in winter with a very deep coat of wool, through which his quills and long coarse hairs project but a short distance. Further south these animals have little or no wool, and in the hot weather I have occasionally seen them entirely destitute of both hair and quills, their naked black skins resembling that of a Chinese dog.

DISAPPEARANCE OF THE MAMMOTH FROM EUROPE AND AMERICA

The mammoth lived in Europe before the Glacial period, and he probably had a wider range in the same continent after that epoch. His final extinction in this region may have been due to human agency. As population increased and the forests became traversed in all directions by highways, and after wide spaces had been cleared by different races of men, the mammoths would find it difficult to maintain their footing. They do not appear to have ranged into Norway or to the southward of the Pyrenees and were very rare in Scotland and Ireland. The geographical boundaries at that time of certain kinds of trees which they preferred for food may have been the cause of thus limiting their distribution.

In North America the last of the mammoths may have been killed off by the aborigines. There is evidence that they hunted these creatures, and no doubt they did so for food. Such a large animal would always be a tempting object of the chase to a people depending for subsistence almost entirely on the product of their hunt. When we see that a few years of shooting by foreign sportsmen in Africa has reduced the elephants of that great continent to a mere fraction of their former numbers, it is not unreasonable to suppose that systematic hunting by the North American Indians throughout many centuries would finally exterminate the mammoth on this continent.

Habitat of the Mammoth inferred from the Form of his Tusks

In both Asia and North America the mammoth probably preferred the open barren-lands or tundras to the thick woods, and in this connection the occurrence of its remains in Alaska, the Yukon and McKenzie River region, in the far northwest of Canada, and on the east coast of Hudson bay is of much interest. The great length and the complete curve of the tusks of these animals show that they were only fitted for traveling in such regions or in very open woods. They would be able to make little or no progress through the thick coniferous forests of Siberia or Canada.

In 1884 I observed on Nottingham island, in Hudson strait, a curious fact bearing on this question in connection with the antlers of the reindeer. On the mainland, where these deer may require to traverse the thick forest in some part of their migrations, their antlers, although much larger and longer than those of the woodland reindeer or caribou, are straight at the tips and of such a form as to be readily dragged through the branches of trees; but on the large island referred to there are no trees of any kind and the antlers of the deer are more spreading, while the tines are strongly curved or hooked. These peculiarities may be merely ornamental or they may be of service to the animal in some other way, but it would be impossible for him to get through any forest. The peculiarities of the tusks of the mammoth, which have been already referred to, would not only prevent the creature from traveling in thick woods, but they would also render the tusks useless for digging up trees, which is the principal use to which both African and Indian elephants put their straighter tusks. These characters would also indicate that the mammoth was adapted only for living where it was not necessary to dig at the roots of trees and to pull them down, but in some region where he might obtain all the brush he required, as he could on the extensive northern plains of both continents in summer, as well as among the small branchy trees at the edge of the forests in winter. The fact that in this habitat the ground would be frozen for the greater part of the year is another reason why he would not use his tusks for digging.

THE OCCURRENCE OF MASTODON REMAINS NEAR THE SOUTHERN EXTREMITY OF JAMES BAY

When at moose factory, at the southwestern extremity of James bay, in the autumn of 1877, I was presented by Mr S. K. Parson, the chief factor in charge, with a very perfect tooth of a mastodon, which had been obtained shortly before my visit in the bed of the Moose river, at its first bend below the junction of the Missinaibi and Mattagami to form this trunk stream.   The locality is 46 miles in a straight line southwestward from moose factory and has an elevation of about 150 feet above the sea.   In the middle of summer of that year, the stream was very low, and an Indian passing down in his canoe happened to see a very large bone, which turned out to be a mastodon's jaw, lying in the shallow water. Setting it on end beside his canoe, he chopped out this tooth with his hatchet, and then allowed the jaw to drop back into the river.   The molar is of a medium size and is very well preserved.   It has nine conical points or tubercles, all of which are entirely covered with enamel. At the same time that I obtained this tooth I was informed by Mr Parson that some years previously a party of Indians had found some large bones in the bed of the Abitibi river, between the lowest on Sextant rapids and its junction with Moose river, which occurs at 18 miles above Moose factory.   From the description, I judged them at the time to be elephantine remains.   The Sextant rapids are at the upper end of the first stretch of the Abitibi river after leaving its mouth.   The superficial deposits are of the same character along this stretch as on the main Moose river for many miles above and below the junction of the two streams.

THE PLEISTOCENE DEPOSITS SOUTHWEST OF JAMES BAY

In order the better to surmise the conditions and the geological time of the existence of the mastodon in the region around the southern extremity of James bay, I should here give a short description of the Pleistocene deposits and of the general character of the district.

Quaternary clays containing recent marine shells extend as far, at least, as the Sextant rapids, and fragments of lignite washed out of these deposits were observed along the margins of the river in this section.

The upward general course of Moose river, which is continued by the

Missinaibi, from Moose factory,* which stands about 7 miles in from the mouth, is southwest for 127 miles to Round bay, at the foot of the Archæan plateau, where it turns south. The rate of rise in the river-bed from the head of tide, 9 miles above Moose factory, to this point is estimated to be between 3 and 4 feet to the mile, which would make the elevation at the end of this distance between 400 and 500 † feet above sealevel. In this interval the river flows with a pretty uniform and rather swift current, interrupted by stony rapids here and there. The banks consist exclusively of till and stratified drift, resting directly upon Silurian and Devonian limestones, which slope almost imperceptibly toward the bay or at only about the same rate as the river-bed itself.

The name Moose river belongs properly to only the trunk stream below the junction of the Missinaibi, or western branch, with the Mattagami, or central branch, at the above mentioned 46 miles from the factory; but as the traveled route to the Canadian Pacific railway follows the former branch, it is sometimes referred to under the same name. The general height of the cut-banks of the river increases from 10 or 20 feet at Moose factory to about 140 feet at the end of the 127 miles referred to, but often for considerable distances the country is low on one side or the other, and sometimes on both sides. The usual height of the cut-banks is from 30 to 50 or 60 feet. Their lower parts appear to be composed, in most cases, of blue clay, sometimes soft and sticky, with or without rounded pebbles. The central stratum, which constitutes the major portion of the banks, consists of bluish-gray and drab clays, with boulders and pebbles. The marine shells washed out of the banks appear to be derived from beds of pebbly drab clays associated with this division. The upper portions of the deposits consist of beds of gravel and sand, with brownish gravelly and bouldery earth towards the top.

The clays near the mouth of the river contain upward of a dozen species of marine shells, some of which indicate tolerably deep water; but in ascending the stream the deep-water species disappear, and at last only the shore and the shallow-water shells remain, such as *Saxicava rugosa*, *Macoma calcarea*, *M. grœnlandica*, *Mytilus edulis*, and *Mya truncata*. Some of these shells were found as far as Round bay, at the extremity of the long southwestward stretch of the river above referred to. Marine shells were also observed along the Mattagami from its mouth up to the foot of the Grand rapids, a distance of 39 miles, and at an altitude which is probably only a little lower than that attained by these shells on the Missinaibi. The species last seen on the Missinaibi were *Saxicava rugosa* and

---

* Factory, the residence of a chief factor or agent.
† In my Geological Survey Report for 1877, p. 7 C, a lower estimated elevation was mentioned, but subsequent observations render it probable that this is more nearly correct.

Macoma fragilis (*Tellina grœnlandica*). These, with a *Leda*, were also the species observed at the highest localities on the Mattagami. Along the Albany river, which flows into the western side of James bay, and also on its great southern branch, the Kenogami, the banks, as well as the deposits of which they are composed, are similar to those of Moose river and its branches. I estimated the elevation of the highest and most inland locality at which I found marine shells on the Kenogami to be 450 feet above the sea.* The Attawapiskat is the largest river flowing into James bay north of the Albany. I surveyed this stream for upward of 300 miles from the sea, and although it flows through a level country and has low banks, I did not detect marine shells at any great distance from the head of tide.

Although the existence of lignite *in situ* in the superficial deposits of the Albany and Abitibi rivers may be inferred from the occurrence of loose pieces of it along their shores, beds of this substance have as yet been noticed only on the Kenogami and the Missinaibi. On the former stream it was found in the bottom of an old channel excavated in the till and again filled up by boulder clay.† This bed contained sticks of coniferous woods and of the canoe birch, but no animal remains were detected in it.

Along the Missinaibi, beds of lignite were seen at a number of places all the way from the foot of the Archean plateau to the junction of the Mattagami. The first of these was in the west bank of a southern branch called Coal brook, three-quarters of a mile from the main river. This bed is three feet or more in thickness, is underlaid by soft sticky blue clay, and overlaid by about 70 feet of till, full of small pebbles, passing into gravel at the top. This lignite contains a little iron pyrites, and much of it retains a distinct woody character. Some of the flattened trunks embedded in it are two feet in diameter.

" On the south side of the river, at nineteen miles below Coal brook or two miles above Woodpecker island, a horizontal seam of lignite was found in the midst of a bank of till 125 feet high. It is from 1½ to 2½ feet thick, and is made up principally of sticks and rushes. Below the lignite are 80 feet of yellow-weathering gray clay and above it 45 feet of blue clay. Both varieties of clay are full of pebbles, and they also hold some striated boulders of Laurentian gneiss, Huronian schists, and unaltered Devonian limestone.

"At three miles below Woodpecker island, or nine miles above the mouth of Opazatika river, another bed of lignite occurs in the bank on the same side. . It is six feet thick, but diminishes to the eastward, and is of a shaly character, being made up of laminæ of moss and sticks. Immediately beneath the lignite is a layer

---

*Geological Survey Report for 1871-'72, p. 112.
†Geological Survey Report for 1871-'72.

one foot thick of irregularly mingled clay and spots of impure lignite. Next below this are 40 feet of unstratified drift, full of small pebbles, under which are a few feet of stratified yellowish sand and gravel. Resting upon the lignite are five feet of hard lead-colored clay, with seams and spots of a yellow color and layers of red, gray, drab, and buff. Above all and forming the top of the bank, 65 feet high, are 10 feet of hard drab clay, with striated pebbles and small boulders and holding rather large valves of *Saxicava rugosa*, *Macoma calcarea* (*Tellina proxima*), and *Mya truncata*. Small seams of lignite were seen in two places in the bank on the same side at and again half a mile below the foot of a rapid which occurs about six miles above the Opazatika.

"In the interval between one and two miles above this stream the whole bed of the river appears to be underlaid by lignite. When sounded with a heavy pole, it has an elastic feel and gives off large volumes of gas, which may also be seen at any time bubbling up spontaneously here and there all along this part of the river. This phenomenon has been observed by the Indians from time immemorial, and the locality has received the name of 'The Bubbling Water.'" *

At the foot of the long portage on Missinaibi river, which is four and a half miles within the Archean border, or that distance south of Round bay, at the head of the long southwestward stretch above described, there is a considerable thickness of fine silt in thin layers, with moss and remains of fresh-water marsh plants between most of them. The mean height of the deposit is about 90 feet above the level of the highest occurrence of the marine shells before mentioned, or probably about 550 feet above the sea. At the time of the postglacial submergence this deposit may have been forming at what was then the mouth of the Missinaibi river, while the whole of the Paleozoic plain between it and James bay was covered by the sea.

The deposits which have been described cover a very extensive district, namely, the low country embraced by a semicircular curve in the great Archean plateau, extending 200 miles southwest from James bay. This tract is all underlaid by the nearly horizontal Silurian and Devonian strata already mentioned. These rocks also appear to form the floor of the bay itself, which is 300 miles long and 150 miles wide.

The lignite beds above described probably all belong to basins of limited extent. The one which has been referred to as occurring in the bottom of a drift-filled valley which had been excavated in older till on the Kenogami river, and also most of the beds along the Missinaibi, are of interglacial age. The seam which has been mentioned as lying beneath a thick stratum of till on Coal brook may be of preglacial age, in which case the blue clay below it would also be preglacial. Some of the higher beds of impure lignite further down the Missinaibi may be post-

* Geological Survey Report for 1877-'78, p. 4 C.

glacial. The stratified deposits with which the lignites are associated are, in part at least, marine, as proved by the shells which they contain.

PERIOD WHEN THE MASTODON LIVED HERE

The mastodon's jaw described above, having been found loose in the bed of the river, may have been washed out of these banks and thus be of either interglacial or postglacial date; but it had not suffered any wear, the tooth being fresh and perfect, and it shows no sign of abrasion. It has been mentioned that lignite occurs *in situ* in the bed of the river where the jaw with this tooth was found.

This relic of the mastodon may belong to a very recent period, perhaps to a time subsequent to the excavation of the river channel out of these lignite-bearing clays, sands, and gravels. Its most ancient possible date would be subsequent to that of the lignite bed on which it rested.

MASTODON REMAINS FOUND IN MANITOBA

Some years previous to 1853 parts of the skeleton of a large mammal, which afterwards proved to belong to a mastodon, were found by Indians in the bottom of the valley of Shell river, at its junction with its east branch. This stream is itself an eastern branch of the Assiniboine, and it takes its rise in the high ground to the west of lake Winnipegosis. The river-flat at the spot where the mastodon remains were found has an elevation of 2,050 feet above the sea, according to Mr J. B. Tyrrell.[*] The scapulæ were the only portions of this skeleton which reached England, and they were examined by Sir John Richardson, who at first gave the species to which they belonged the provisional name of *Elephas rupertianus*, but afterwards, on making critical comparisons with the bones of other fossil elephants, he wrote :

"The probability, therefore, is that the Swan river (*i. e.,* Swan River district) bones belonged to the *Mastodon giganteus*, and that the range of that species must be extended northward in Rupert's land to the fifty-second parallel of latitude, while the provisional geographical designation of *Elephas rupertianus* must be expunged." [†]

DEPOSITS IN WHICH THE MASTODON BONES WERE FOUND

In 1874 I examined the Assiniboine river all the way from Fort Pelly to Fort Ellice and also a part of Shell river, and sent my assistant, Dr

---

* Report of the Geological Survey of Canada for 1890-'91, p. 129 E.
† Zoology of the Voyage of H. M. S. *Herald* during the years 1845-'51. London. Lovell Reeve, 1854, pp. 101, 141.

J. W. Spencer, further up the latter stream. As the age of the deposit in which these mastodon remains were found and the nature of the surrounding country were important in connection with the discovery, I quote Dr Spencer's report to me on these points : *

" The valley of the Assiniboine adjacent to that of the Shell river is about a mile wide and some 200 feet deep. The alluvial flat at the bottom of the valley is three-quarters of a mile wide, and the banks rise steeply on either side. Through this level flat the river pursues a meandering course from side to side, occasionally leaping a small rapid caused by the obstruction of Laurentian boulders. Twenty miles further up, the valley is nearly 3 miles wide, but at this place, in the bottom and following the valley longitudinally, there are 4 or 5 series of hills, rising irregularly one above the other, till the highest reaches nearly to the level of the plain above. Between these hills there are small deep valleys. The western bank is often strewn with gravel and boulders, while the flats below are nearly free from them, excepting in places along the bed of the river. The sides of the valley are often deeply gorged, but the ravines do not extend to any great distance back from the valley. Many of them appear to have been cut out by the waters from springs. These springs usually hold a considerable quantity of iron in solution, and I observed several places where yellow ochre was being deposited around them. In several localities on the banks of the Assiniboine extensive landslides are to be met with, sometimes showing stratified deposits of clay or sand. The general course of the Assiniboine river at the influx of the Shell river is nearly south, but above this it has a more westerly direction.

" I explored the Shell River valley upward for 30 miles, and Mr Hagar continued the exploration for 10 miles further. Along the upper part of this distance the country on either side has usually a rolling prairie character, while in the lower portion the river flows in a valley nearly as wide and deep as that of the Assiniboine. The general course is nearly from the north. At the bends of the valley the river usually winds its way to the outer side, and on the inner side of the curve there is left a terrace or series of terraces rising from the alluvial flat to the plain above. The country is generally wooded, except here and there where fires have swept over small areas. The Shell river is much more rapid than the Assiniboine, and the sides of the valley are much more deeply gorged than those of the latter river. At the landslides along the Shell river I observed a few stratified deposits, but they generally showed only a heterogeneous mixture of gravelly earth, with boulders. . . .

" If the rivers which now flow through them have excavated these valleys, the former must be of great antiquity. The valleys are yearly becoming larger by the spring floods bearing away great quantities of material. Everywhere along the river banks there are evidences of former landslides. . . . The deposits of the Shell River valley frequently consist of irregular beds of clay, with boulders, while along the alluvial flat of the Assiniboine they consist of regularly stratified clays. The summits and sides of the banks of both streams are generally covered with boulders.

" In the more recent deposits of the Shell River valley an Indian is said to have found, a few years ago, some large bones, which were at the time sent to Fort

* Report of the Geological Survey of Canada for 1874-'75, pp. 58 and 63.

Ellice and afterwards to England. These remains were described to me by a man who had seen them, and also the place whence they came. They appear to have been large enough to have belonged to *Elephas.*"   .   .   .

In the summer of 1887 Mr J. B. Tyrrell, of the Geological Survey, made a further examination of Shell river, the results of which are published in the Report of the Department for 1890–'91. Referring to these bones, he says :

"The Indians allege that at this point (the junction of the north and east branches) huge bones were found at the bottom of a landslide and were brought to the officer in charge at Fort Pelly, by whom they were forwarded to England. Hon. W. J. Christie, of Brockville, Ontario, who was in charge of Fort Pelly at the time, informs me that the bones were shoulder-blades, and that in 1853, some years after the bones were brought in, he visited the place 'and examined the spot carefully where the blades were taken out of the river at low water. A landslip had occurred from the bank and carried the bones into the river. I found, from cross-questioning my guide, that the Indians had collected the bones and burnt them on the bank, from superstition, and buried what would not burn. I examined the spot where they had buried the bones, but what remained crumbled to pieces when touched.' " *

When traveling in the Northwest territories in 1873 I was informed that large bones, supposed to be those of elephants, had been found at Sand Hill lake, near the elbow of the South Saskatchewan river, and also on the surface of the ground on White Mud river, a small tributary of the Missouri on the west side of the Cypress hills,† but I have never been able to verify these reports.

Discoveries of Elephant Remains in other Parts of Canada

Numerous discoveries of remains of both mammoths and mastodons have been made at various times during the past seventy years in the province of Ontario, but with one exception they all occurred in the district south of a straight line drawn west from Toronto to the outlet of lake Huron. The exception was the finding of the greater part of the skeleton of a very large mammoth in a swamp on lot 9, range VII, of the township of Amaranth, county of Wellington, at about 50 miles northwest of Toronto. A tusk found with this skeleton was reported to measure 8 inches in diameter and 14 feet in length.

In Ontario the remains of the above animals have always been found under similar conditions and in very recent deposits. In a few instances they have been met with in gravel and sand. At Burlington heights, at

† Report of the Geological Survey of Canada for 1873, pp. 73, 74.
* Op. cit., p. 129 E.

the western extremity of lake Ontario, the bones and tusks of a large mammoth were found in 1848 in an ancient beach deposit, 37 feet below the surface and at an elevation of 70 feet above the lake; but in the majority of cases such remains have been discovered just beneath the surface of the ground in cutting ditches to drain swamps, in shell marl left by the drying up of small shallow lakes, or under thin coverings of superficial loam, silt, etcetera. In the province of Quebec no traces of fossil elephants have yet been recorded, as far as the writer is aware.

In the maritime provinces the only discoveries of such relics yet made consist of some mastodon bones which were found in banks of sand and gravel in the valleys of Middle and Baddeck rivers, in the central part of the island of Cape Breton, which forms part of Nova Scotia. The elevation in each case did not extend 50 feet above the sea, and the two localities are less than 20 miles apart.

NOTE.—Since the foregoing paper was in print the writer has received the June number of the Scottish Geographical Magazine, containing Professor James Geikie's thoughtful article on "The Tundras and Steppes of Prehistoric Europe," in which he makes some remarks that are of interest in confirmation of the manner in which I have supposed that some of the mammoths of Siberia had perished. He says: "We have seen how in existing tundras and steppes the semi-domesticated and wild animals of these regions are now and again overwhelmed in storms and smothered in snow. Now, similar catastrophies must have happened again and again in the tundras and steppes of prehistoric times."